River Keep Flowing,

Oh Well

River Keep Flowing, Oh Well

IBSN: 978-0-578-67885-6

Cover design by Jerome Littles.
Book design by Jerome Littles.

Printed in the United States

First printing edition 2020.

jerlittles@gmail.com

chelseandrome.com

For Roman Wilde. Always believe in yourself and never allow outside views to determine your path.

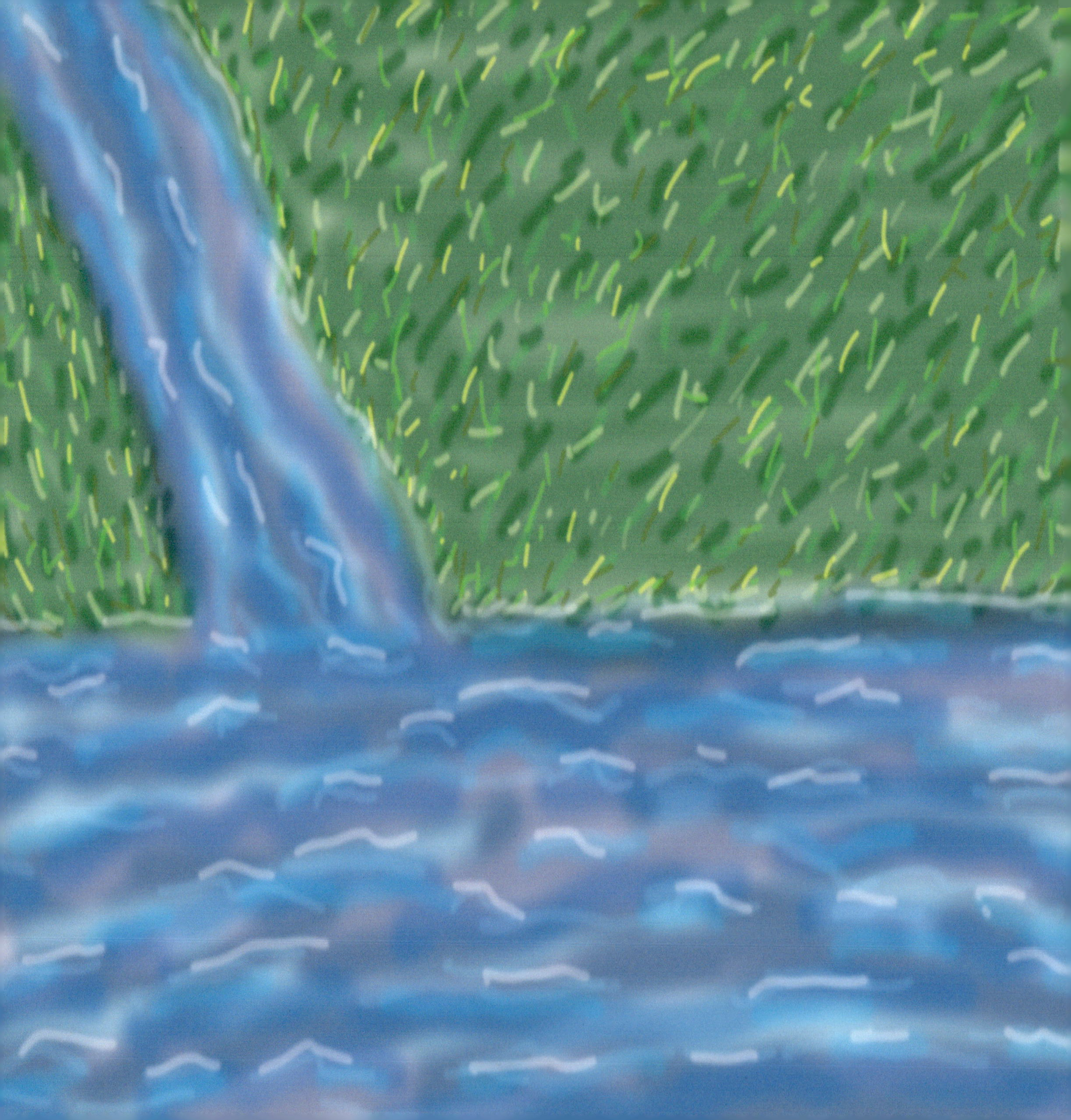

River was always flowing
and growing,
constantly moving towards
different places and spaces.

But, Well was stagnant - staying in one place and depending on Aquifer to be filled.

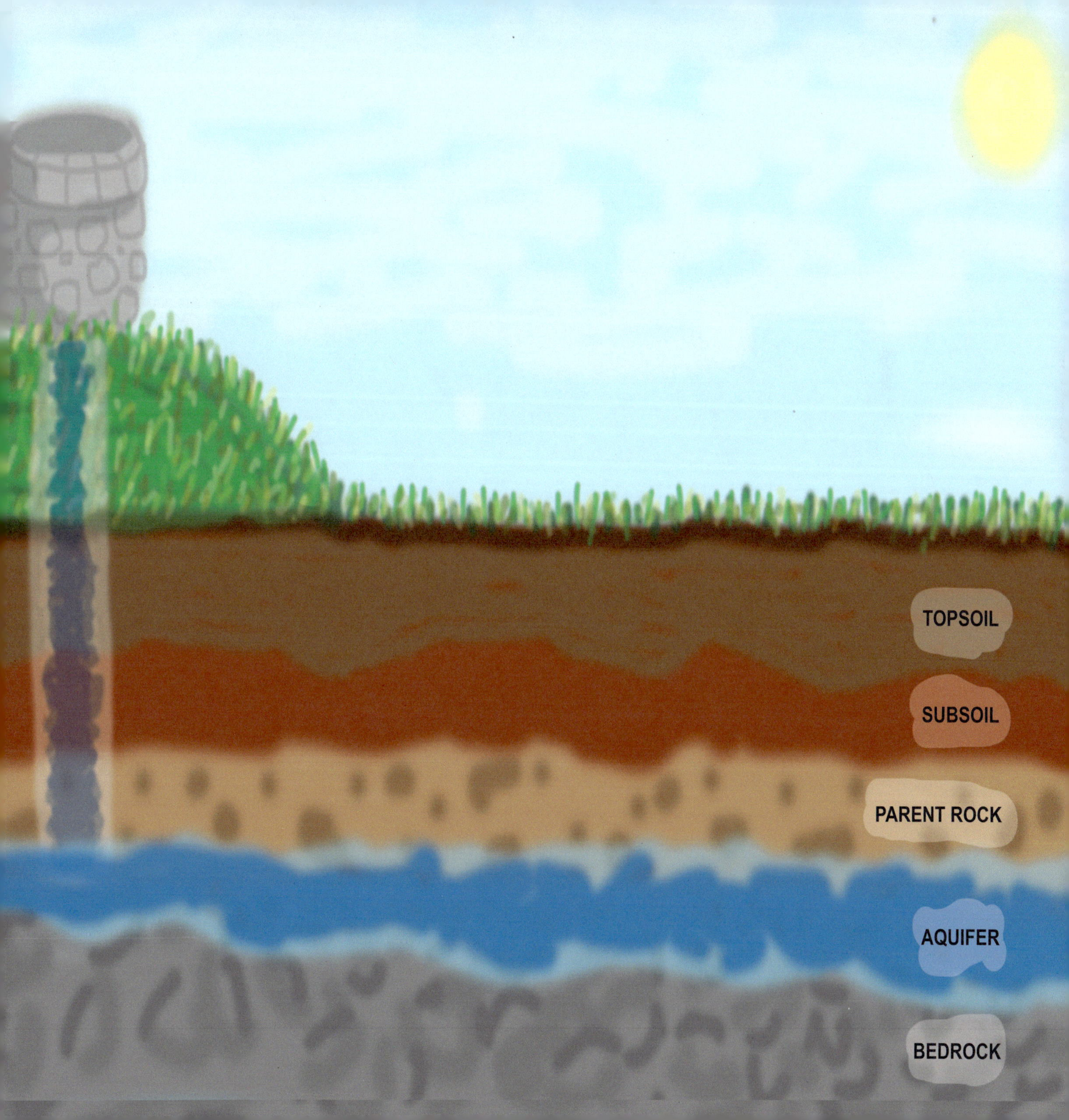
TOPSOIL
SUBSOIL
PARENT ROCK
AQUIFER
BEDROCK

Well's life was great -
having nothing to do
but chill on a hill.

So River thought...

River watched everyone gathered around Well, happy and skipping, while filling water in their pales.

River felt sad by all the attention
Well got;
only having to do one thing:
just sit in one spot.

Ocean came to River saying;
“There are consequences that come with
not putting forth effort.
I’ve seen you looking up there at Well
sitting high and being glorified.
That does seem like the good life!
Not having to work and
getting all that attention.”

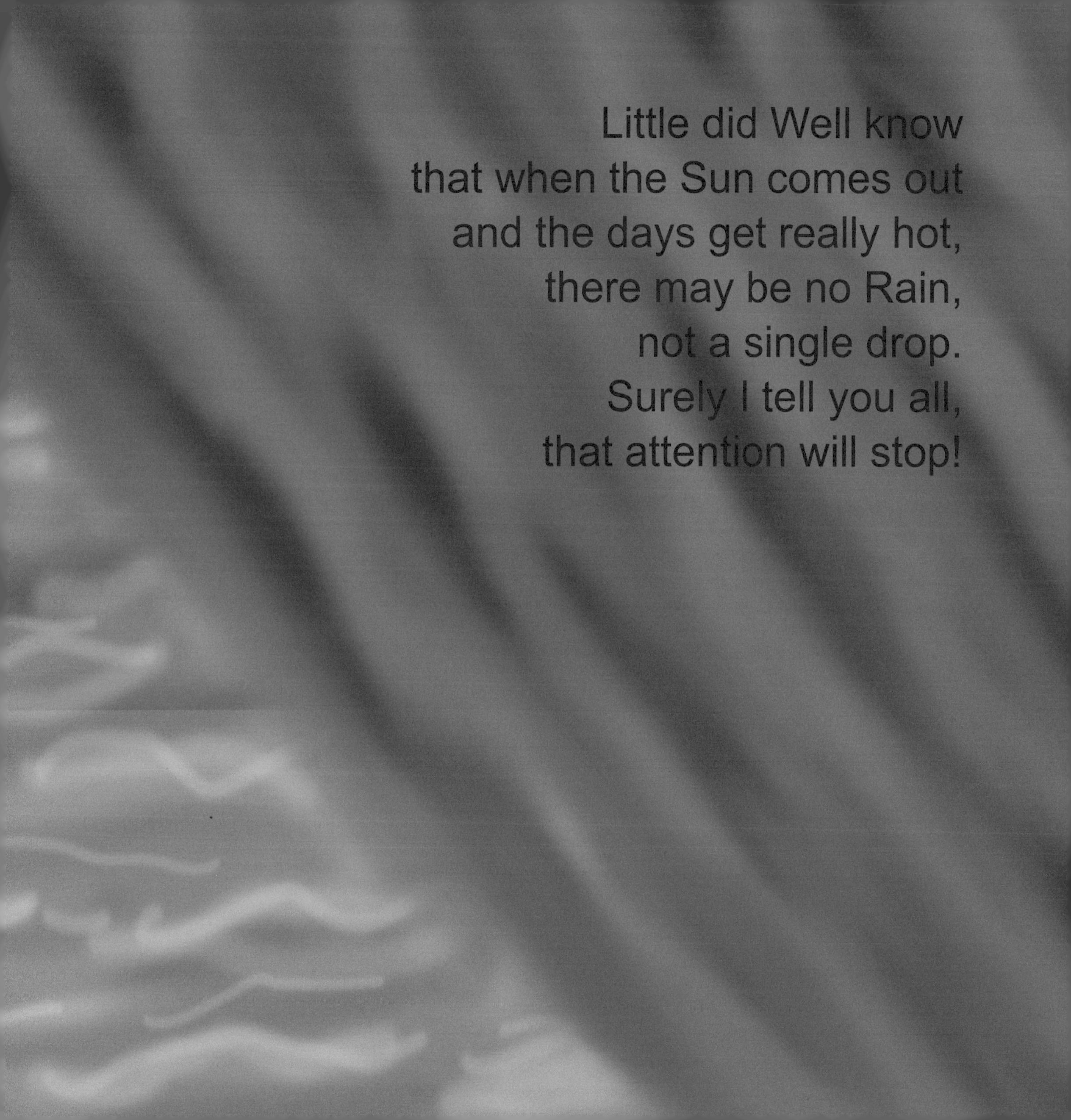

Little did Well know
that when the Sun comes out
and the days get really hot,
there may be no Rain,
not a single drop.
Surely I tell you all,
that attention will stop!

Soon, Well will run dry
from the Sun staying high in the sky
and everyone drawing Well dry.

River your life is great too!
Your work is not in vain,
because there would be no Well
without the Rain!

DEFINITION BANK

RIVER - a large stream of water that flows to the ocean

WATER WELL - a structure created in or above the ground by digging or drilling to access ground water from aquifers

OCEAN - a larger body of salt water that covers most of the earth's surface

TOPSOIL - great for plants and other organisms to live

SUBSOIL - has a lot of minerals made from the topsoil

PARENT ROCK - the original rocks that sunk below the earth's surface

AQUIFER - a body of loose rocks which can contain or carry groundwater. There are two types: confined and unconfined. Well is filled by an unconfined aquifer

BEDROCK - a variety of different types of rocks group together found below the aquifer

GROUNDWATER - water held underground in the soil or in pores and crevices in rock

RAIN - water from rivers, lakes, streams, oceans evaporates into the air that is heated up by the sun and then cools down

EVAPORATION - when water is changed into a gas

PRECIPITATION - when the evaporated water in to the sky drops back down to the earth

ESTUARY - the place where the river and the ocean meet

FOSSIL - the parts that are left after animals pass away and are found in the earth after the water is gone

Can you find the images of...

The layers of the soil?

Fresh water?

Salt water?

Groundwater?

The water bank?

Rain clouds?

Evaporation?

Precipitation?

The estuary?

A fossil?

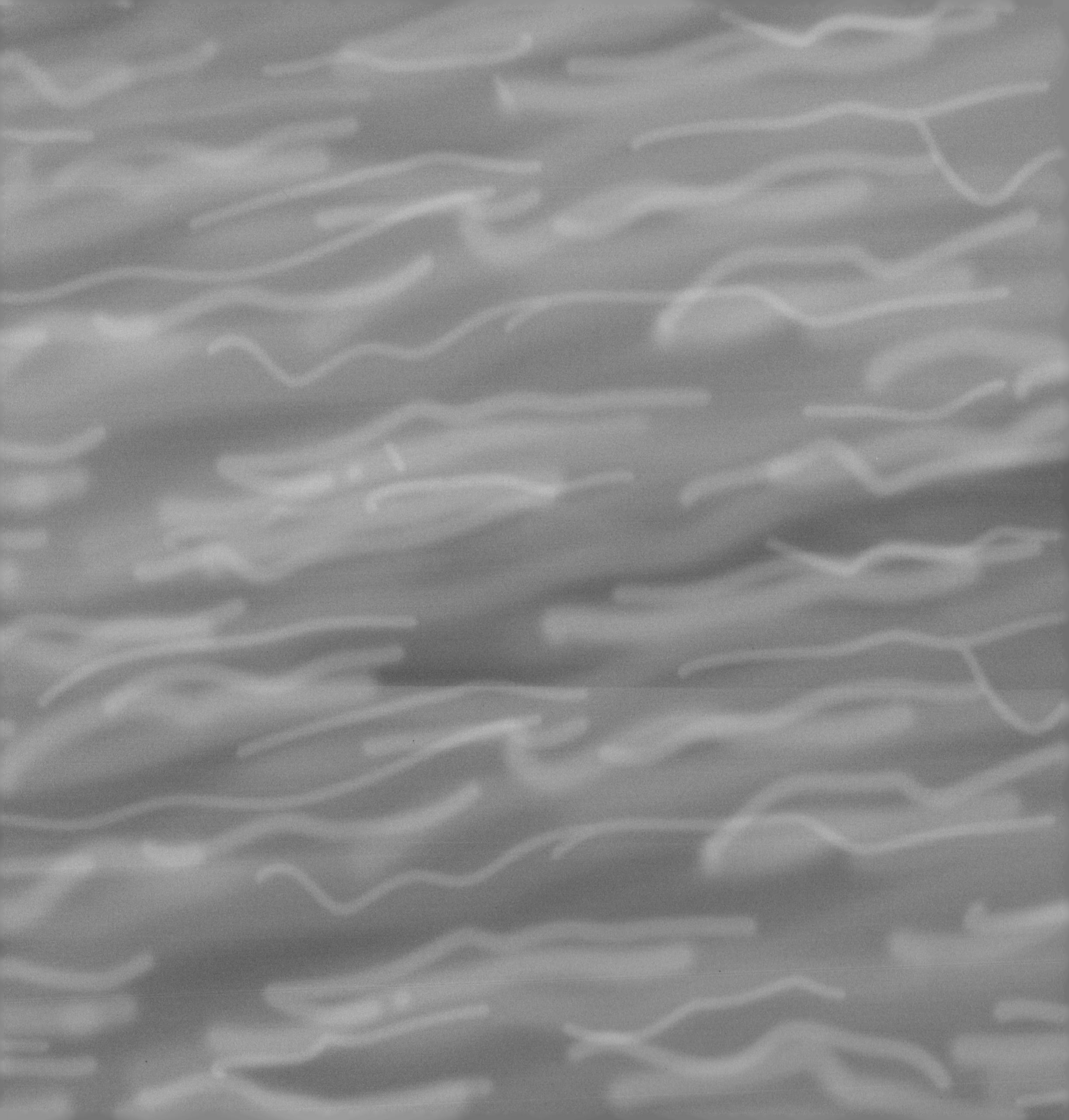

Made in the USA
Monee, IL
05 November 2026

67383138R00017